BACKYARD SCIENCE

TERRIFIC TREES

Harold Morris

TABLE OF CONTENTS

A Crabtree Seedlings Book

CRABTREE
Publishing Company
www.crabtreebooks.com

2

TREES

Trees are living things. Trees grow in most places on Earth.

Trees provide food and shelter for other living things.

Trees have a life cycle.

Each acorn contains a seed.

Acorns grow on oak trees. When acorns fall to the ground some grow into new trees.

Many trees can live 40 or 50 years. Some can live hundreds, or even thousands, of years!

Bristlecone pine trees are among the oldest trees on Earth.

How do we know
the age of a tree?
We count its growth rings.

each ring = one year

How old was this tree?

Growth Ring

Some trees lose their leaves in the fall. These trees are **deciduous** trees.

Evergreen trees do not lose their leaves. They stay green all year.

Most evergreens have needles, not leaves.

Some trees look different in each **season**.

Winter

Spring

Summer

Fall

Like all plants, trees get energy through **photosynthesis**.

During photosynthesis, trees take in sunlight, water, and carbon dioxide. They release **oxygen** into the air.

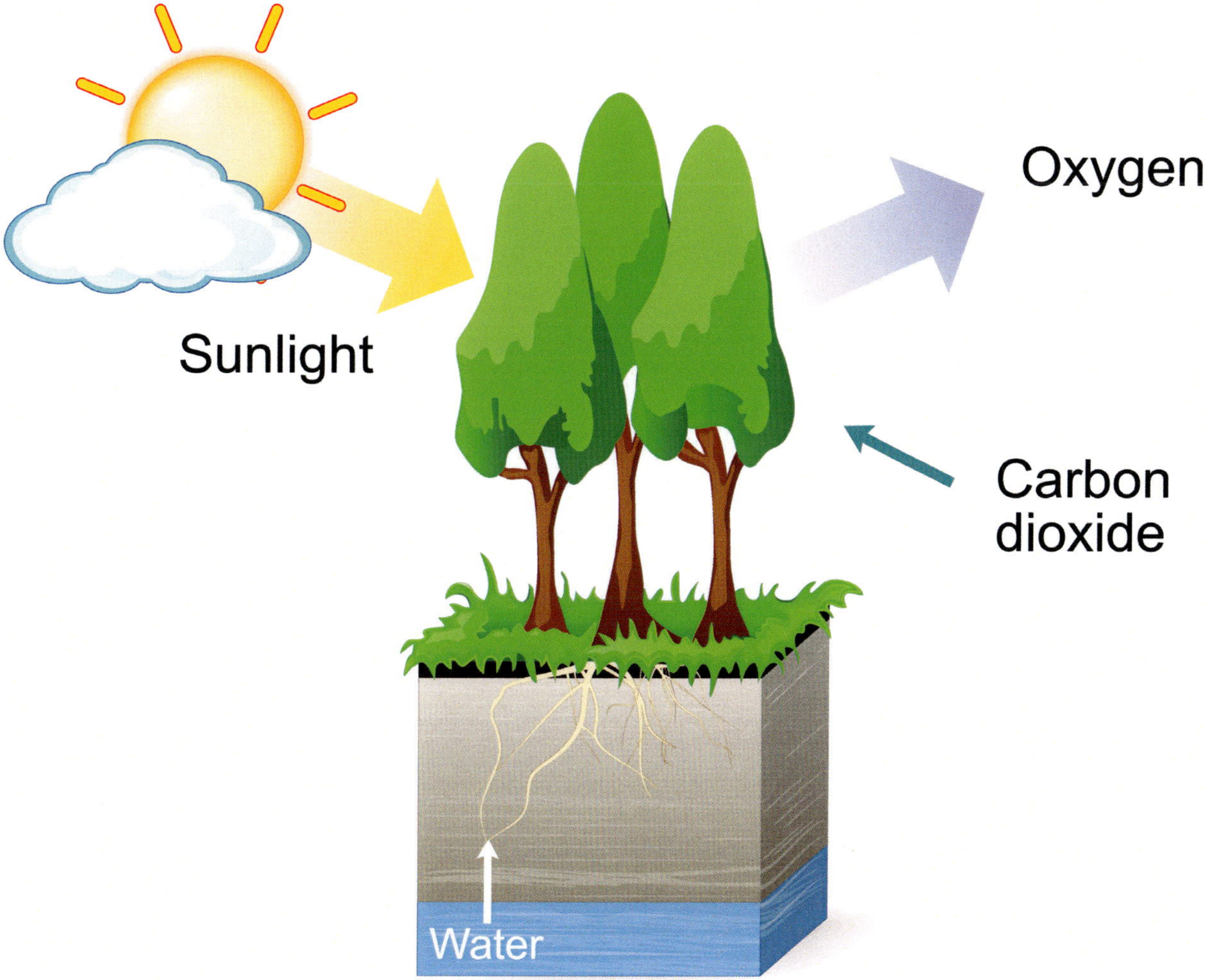

Be a backyard scientist!

Study the trees in your backyard or in a park. Take notes. Are they deciduous or evergreen? What shape are the leaves? Do any animals live in them?

Glossary

deciduous (di-SIJ-oo-uhss): Deciduous trees lose their leaves every year.

evergreen (EV-ur-green): Evergreen trees stay green all year. They do not lose their leaves or needles.

oxygen (OK-suh-juhn): Oxygen is a colorless gas found in the air. Humans and animals need oxygen to breathe.

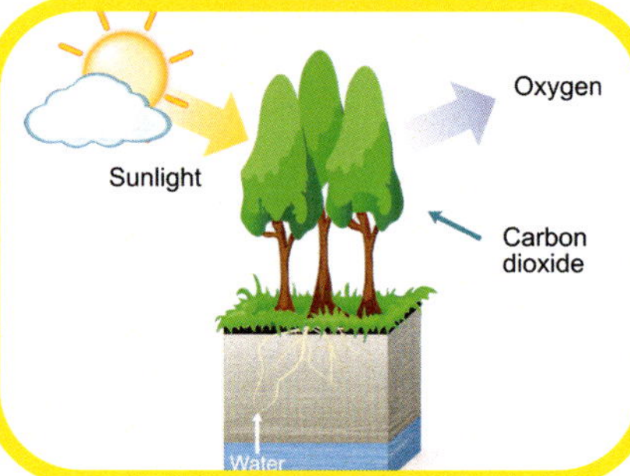

photosynthesis (foh-toh-SIN-thuh-siss): Photosynthesis is how plants make their food using sunlight, water, and carbon dioxide.

sapling (SAP-ling): A sapling is a young tree.

season (SEE-zuhn): A season is each of the following four parts of the year: spring, summer, fall, and winter.

Index

School-to-Home Support for Caregivers and Teachers

This book helps children grow by letting them practice reading. Here are a few guiding questions to help the reader build his or her comprehension skills. Possible answers appear here in red.

Before Reading

- **What do I think this book is about?** I think this book is about all different kinds of trees. I think this book is about how trees change during the year.
- **What do I want to learn about this topic?** I want to learn more about trees in my neighborhood. I want to learn why some trees lose their leaves and others don't.

During Reading

- **I wonder why...** I wonder why some trees grow and live for hundreds of years and others only live for 50 years. I wonder why evergreen trees do not lose their leaves or needles and stay green all year.
- **What have I learned so far?** I have learned that photosynthesis is how plants and trees make their food using sunlight, water, and carbon dioxide. I have learned that trees have a life cycle and that they grow from seeds.

After Reading

- **What details did I learn about this topic?** I have learned that trees provide food and shelter for other living beings. I have learned that each year a tree is alive it gains a new growth ring, and that we can count the growth rings to find out how old a tree is.
- **Read the book again and look for the glossary words.** I see the word *sapling* on page 6, and the word *evergreen* on page 14. The other glossary words are found on pages 22 and 23.

Library and Archives Canada Cataloguing in Publication

CIP available at Library and Archives Canada

Library of Congress Cataloging-in-Publication Data

CIP available at Library of Congress

Crabtree Publishing Company
www.crabtreebooks.com 1–800–387–7650

Written by: Harold Morris
Production coordinator and Prepress technician: Tammy McGarr
Print coordinator: Katherine Berti

Print book version produced jointly with Blue Door Education in 2022

Printed in Canada/062022/CPC20220603

PHOTO CREDITS:
istock.com, shutterstock.com, COVER: main photo © istock.com /Artazum, child © istock.com /wavebreakmedia, fence illustration © istock.com /ecco; PG2-3: Riekkinen. PG4-5; istock.com /fpwing, momnoi. PG6-7; istock.com fotofermer, dionisvero, AndriiM, Zerbor. CLopetz. stevanovicigor. PG8-9;istock.com/SharonDay. PG10-11; emer1940. Pg12-13; istock.com /puflic_senior PG14-15; istock.com /zlikovec, Azure-Dragon. PG16-17; istock.com /LiliGraphie. PG18-19; istock.com /ttsz., max-kegfire. Pg20-21; istock.com / fpwing, Leonid Andronov. PG22-23; istock.com /Fotofermer, mandritoiu, LongQuattro

Published in the United States
Crabtree Publishing
347 Fifth Ave.
Suite 1402-145
New York, NY 10016

Published in Canada
Crabtree Publishing
616 Welland Ave.
St. Catharines, Ontario
L2M 5V6